Living Green

Respecting Nature

By Meg Gaertner

level 2 little blue readers

www.littlebluehousebooks.com

Copyright © 2023 by Little Blue House, Mendota Heights, MN 55120. All rights reserved. No part of this book may be reproduced or utilized in any form or by any means without written permission from the publisher.

Little Blue House is distributed by North Star Editions:
sales@northstareditions.com | 888-417-0195

Produced for Little Blue House by Red Line Editorial.

Photographs ©: Shutterstock Images, cover, 4, 7, 9 (top), 9 (bottom), 11, 12, 17, 18, 21, 22–23, 24 (top left), 24 (bottom left), 24 (bottom right); iStockphoto, 14–15, 24 (top right)

Library of Congress Control Number: 2022901877

ISBN
978-1-64619-598-5 (hardcover)
978-1-64619-625-8 (paperback)
978-1-64619-676-0 (ebook pdf)
978-1-64619-652-4 (hosted ebook)

Printed in the United States of America
Mankato, MN
082022

About the Author

Meg Gaertner enjoys reading, writing, dancing, and being outside. She lives in Minnesota.

Table of Contents

tent

At Parks

My family respects nature in the park.

We spend many days outside.

We go camping in a tent.

Sometimes we walk through the woods. We stay on the trails. We point to our favorite sights.

woods
trail

I like pretty flowers, and my brother likes singing birds. My mom likes puffy clouds, and my dad likes tall trees.

flower
bird

tree
cloud

We leave the campsite how we found it. We take our things home, but we remember nature's beauty.

campsite

At Home

Nature is all around us.
So, my family respects nature
at home, too.

We plant a garden
and care for our flowers.
They grow and bloom
with pretty colors.

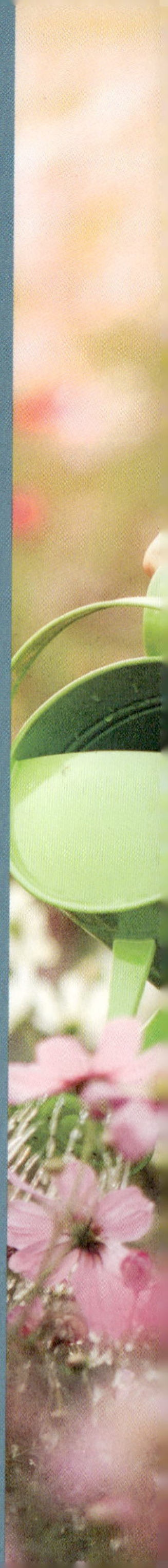

garden

We hang bird feeders.
Birds come every day to eat seeds.
We learn their different names and songs.

bird feeder

trash

Clean and Safe

My family respects Earth and keeps it clean.

We pick up our trash.

We leave wild places alone.

They are animals' homes.

We respect animals and keep them safe. We don’t scare them. Watching from a distance is fun.

My family loves nature.

We work to keep it clean and safe.

Earth stays beautiful.

Everyone can enjoy it.

Glossary

bird feeders

garden

campsite

trail

Index